Tarik Ainane
Hafida Hanine
Ayoub Ainane

Valuing Intellectual Property through Patents

Tarik Ainane
Hafida Hanine
Ayoub Ainane

Valuing Intellectual Property through Patents

Convergence of Innovation and Invention

ScienciaScripts

Imprint
Any brand names and product names mentioned in this book are subject to trademark, brand or patent protection and are trademarks or registered trademarks of their respective holders. The use of brand names, product names, common names, trade names, product descriptions etc. even without a particular marking in this work is in no way to be construed to mean that such names may be regarded as unrestricted in respect of trademark and brand protection legislation and could thus be used by anyone.

Cover image: www.ingimage.com

This book is a translation from the original published under ISBN 978-620-6-72155-0.

Publisher:
Sciencia Scripts
is a trademark of
Dodo Books Indian Ocean Ltd. and OmniScriptum S.R.L publishing group

120 High Road, East Finchley, London, N2 9ED, United Kingdom
Str. Armeneasca 28/1, office 1, Chisinau MD-2012, Republic of Moldova, Europe
Printed at: see last page
ISBN: 978-620-8-09830-8

ENHANCING THE VALUE OF INTELLECTUAL PROPERTY THROUGH PATENTS: CONVERGENCE OF INNOVATION AND INVENTION

PROFESSOR TARIK AINANE, ECOLE SUPERIEURE DE TECHNOLOGIE DE KHENIFRA, UNIVERSITE SULTAN MY SLIMANE, MOROCCO.

PROFESSOR HAFIDA HANINE, BENI MELLAL FACULTY OF SCIENCE AND TECHNOLOGY, SULTAN MY SLIMANE UNIVERSITY, MOROCCO.

PROFESSOR AYOUB AINANE, KHENIFRA COLLEGE OF TECHNOLOGY, SULTAN MY SLIMANE UNIVERSITY, MOROCCO.

PREFACE

In an age that is based on innovation and creativity as the means to economic competitiveness and social development, intellectual property, or to be precise, patents, are gaining enormous importance. Patents can no longer be considered as a mere piece of legal paper; they are a strategic tool that enables an invention to be transformed into a tangible, physical and economically viable asset.

Yet the complexity of regulation, the prohibitive costs of filing, and the need to exercise rights in a timely and targeted manner mean that patent rights are a highly underused and poorly understood protection mechanism.

Valuing Intellectual Property through Patents: Converging Innovation and Invention has been written to fill this gap by providing a comprehensive and in-depth exposition of all issues relevant to patent management. The book is structured in such a way as to be useful to a wide range of readers, including people such as lawyers, engineers, business people, as well as any other learner who wishes to know and learn how the process of valorisation of innovation works.

In the General Introduction, we will lay the foundations for our thinking by giving the reader a definition of the function of the patent in the general context of innovation. We will clarify how the patent can be a tool for economic growth and public policy, while at the same time highlighting the problems associated with its use.

Chapter 1 deals with Basic Concepts. It is here that the reader will have a clear and succinct idea of the key terms involved in intellectual property and patenting. The chapter will elaborate on the notion of patentable invention, prior art, in addition to the prerequisites for patentability, and will clearly explain the importance of these concepts in the valorisation process.

Chapter 2 is on the Legal Regime of Patents, one of the key areas of knowledge relating to the regulatory framework for the protection of inventions. This chapter explains the regulatory provisions relating to the granting, protection and exercise of patents in detail in the light of current international and national law.

The filing of a patent, of which Chapter 3 is the version, is the most important act that any inventor or company can do to protect an invention. The filing requirements will be explained here, what needs to be done to obtain the best possible protection, and the disputes and bureaucratic problems that can arise. This chapter will be your orientation manual for successfully navigating the labyrinth of patent filing. Strategic Patent Management, discussed in Chapter 4, is the essence of value enhancement. It explains how to integrate patents into a company's overall strategy, how to strengthen the company's competitive position through patents, and how to extract maximum value from patents through best practice. We will also discuss approaches, pricing and protection measures required to preserve patent rights. Chapter 5 deals with the management of patent licences and agreements. As soon as a patent is granted, it becomes clear whether there is an opportunity for its owner to exploit it commercially or not. The chapter covers the types of licences, negotiation techniques, and the types of agreements, including licensing agreements, that can be signed to exploit patent rights to the full. We will also discuss the legal aspects of licence agreements and the preventive measures that can be taken to avoid disputes.

Finally, Chapter 6 deals entirely with Valuing Patents for Societal, Political and Economic Reasons. This is where we will open the focus to discuss the wider impacts of patents on society. We will discuss the social innovation made possible by patents, the impacts they have on public policy, and the importance of patents. This chapter will provide an overview of the impacts of patents, going beyond legal protections, and will attempt to show how they can be used as means for the public good.

This book is a pragmatic work as well as a serious meditation on the status of patents in the modern world. By questioning the locus of creativity and invention, it actually inspires the reader to see the patent system more as a visionary than an analytical technology that is capable of stimulating an economy as an engine of strategy and supporting balanced social development. "Valorisation de la Propriété Intellectuelle par le Brevet: Convergence de l'Innovation et de l'Invention" is therefore destined to become a handbook for all those who believe in the power of ideas and for whom transforming abstractions into concepts and tangible things is a way of life. Whether you are an inventor, a lawyer, an entrepreneur, or simply looking to master the details ofinnovation, this book will guide you forward in your work and help you succeed in the tough but promising world of patents.

CONTENTS

INTRODUCTION

Making the most of patents and innovation is an essential process in today's industrial landscape, shaped by a complex interconnection between legislation, operational strategies and technological advances. Patents, in their capacity as legal protection and financial incentive, stimulate innovators while promoting the dissemination of knowledge. The patent filing stage, which is mainly technical and governed by legal laws, registers and subjects the invention to an assessment of its originality, feasibility and usefulness. This process gives rise to a detailed technological roadmap, defining the stages of development, implementation and marketing of the invention, thus laying the foundations for its subsequent exploitation.The industrial exploitation of patents is of crucial importance in transforming patented concepts into marketable products or services. This often requires substantial investment in research and development, as well as marketing strategies to maximise their economic impact. However, the protection of intellectual property rights remains a major challenge, as counterfeiting and infringement of patents can jeopardise the economic benefits of inventors.Current challenges in patent valorisation include efficient management of patent portfolios, appropriate monetisation of inventions and adaptation to emerging technologies. Trends point to the increasing globalisation of patents, requiring a more agile and strategic approach to valuation in order to remain competitive in the ever-changing global marketplace. The outlook envisaged incorporates increased convergence between various public and private industrial sectors, in-depth digitisation of the valorisation processes and a focus on the social and environmental impact of innovations. This vision promotes patents not only as profit-making tools, but also as instruments for solving major social and environmental problems.Government policy varies according to each country's own resources and development strategies, while the issues surrounding the exploitation of patents encompass both national and international regulation and protection. Striking a balance between protecting inventors' rights and disseminating knowledge remains a central challenge in this field. The contexts surrounding patents require a strategic approach, influenced by technological advances, economic dynamics and the legal and social considerations of every part of society. Effective understanding and management is of paramount importance to businesses and innovators in a world of constant change. This book aims to explore the many facets of this complex dynamic, focusing on the convergence between innovation and invention through the prism of patents.

CHAPTER 1
BASIC CONCEPTS

I. Key concepts :

Understanding these key concepts is crucial to understanding the value of patents. Before exploring the basis of "Valuing Intellectual Property through Patents: Convergence of Innovation and Invention", it is necessary to define certain key terms in order to understand the technical system involved.

- **Intellectual property :**

Intellectual property is a set of legal rights granted to creators and inventors to protect their intellectual creations. It encompasses various forms of protection such as patents, copyright, trademarks and industrial designs, aimed at stimulating creativity and innovation while granting inventors exclusive rights to economically exploit their works.

- **Invention:**

An invention is a new and original idea that solves a specific technical problem, thereby initiating the innovation process. It may take the form of a product, process, device or method. To be patentable, an invention must be new, involve an inventive step that is not obvious to an expert in the field concerned, and demonstrate industrial applicability.

- **Innovation:**

Innovation is the creation, adoption and diffusion of new ideas, products, services or practices that add value. It is the translation of an idea into reality, by implementing a new or improved solution to an existing problem. Innovation can take different forms, such as technological, process, product or organisational.

- **Patent :**

A patent is an industrial property title granted by a competent patent office, granting its holder an exclusive right to an invention for a specified period of time. In other words, it gives the inventor the power to prohibit others from making, using or selling his invention without authorisation. Patents are generally granted for new, non-obvious technical inventions with industrial applicability.

II. The importance of technology transfer :

The importance of technology transfer lies in its pivotal role in the innovation process, acting as the basis for transforming inventions into commercial successes. This process represents the progression from a simple idea to the creation of a product or service that meets market requirements. The aim of this section is to highlight the basis of technology transfer in the transition from ideas to commercially and economically viable patents, and to explain how it helps to optimise the economic value of these innovations.

II.1. Transforming the idea into patentable and commercial value:

The motivation behind patentable development lies in the transformation of an idea into a commercial and economic reality, an essential process for innovation players and entrepreneurs. This transition aims to create economic value by obtaining a patent that provides a protection strategy for the invention and exclusivity of exploitation. Patents offer a range of exploitation opportunities: the development of products or services based on the invention, the granting of licences to other companies in return for royalties, or even their transfer to third parties. In addition to their financial value, patents consolidate a company's reputation and attractiveness to investors and business partners. They erect a barrier to market entry, strengthening the position of the holder and contributing to the company's overall growth strategy.

II.2. Technology transfer :

Technological development aims to convert an invention into a technically, economically and commercially viable product or service. This process involves optimising the design, complying with current national and international standards and regulations, assessing the market and selecting appropriate distribution channels. It also incorporates fundamental considerations relating to cultural aspects, regional specificities, environmental contexts and socio-economic impacts.

Entrepreneurial valorisation aims to assign economic value to patentable ideas by transforming them into sources of revenue and profit. Appropriate valorisation can turn patents into valuable assets, both for individual inventors and for small, medium-sized and large companies. Various valuation approaches can be adopted:

• Internal development :

Patent holders can choose to exploit their inventions by creating a dedicated entity or by integrating them into their existing operations. This strategy enables them to benefit directly from their inventions by developing products or services based on them.

• Granting of operating licences :

An alternative for patent holders is to grant user licences to other companies. This enables them to generate income by authorising third parties to use their inventions in return for royalties. These licences represent a stable source of income, while encouraging wider dissemination of inventions on the market.

• Transfer or sale of patents :

In some cases, patent owners may choose to assign or sell their patents to other companies or investors. This can be an attractive option for inventors looking for immediate financial gain or wishing to focus on new innovation opportunities.

III. Closed innovation - open innovation :

Closed innovation and **open innovation** are two distinct approaches to developing new ideas, products or services within companies or organisations. Each has its own characteristics, advantages and challenges.

III.1. Closed Innovation :

III.1.1. Definition:

Closed innovation is a traditional approach where the innovation process takes place exclusively within the company. All stages in the design, development and marketing of new ideas are controlled and carried out internally, without involving external parties.

III.1.2. Features :

• Internal control: All research, development and marketing activities are carried out by the company's employees, under the direct supervision of management.

• Confidentiality: Information relating to innovations in progress is strictly protected, reducing the risk of sensitive information leaking out.

• **In-house R&D:** Research and development (R&D) is mainly carried out within the company, with substantial investment in laboratories, staff and in-house technologies.
• **Exclusivity:** Innovations are considered to be the company's exclusive assets, which can enable it to maintain a unique competitive advantage.

III.1.3. Advantages :

• **Security of ideas :** The risk of theft of ideas or technologies is minimised thanks to the total confidentiality of ongoing projects.
• **Total control:** The company has complete control over the entire innovation process, from idea to market launch.
• **Intellectual property:** The company retains all rights to its innovations, which means it can make exclusive use of patents, trademarks and other forms of intellectual property.

III.1.4. Disadvantages:

• **High cost:** Closed innovation requires significant resources, particularly in terms of finance, infrastructure and manpower.

• **Risk of isolation:** By limiting innovation to internal resources, the company may miss opportunities for new ideas or emerging technologies from outside.
• **Long development times :** The development of new ideas can take longer due to the limited resources available in-house.

III.2. Open Innovation :

III.2.1. Definition:

Open innovation is an approach that encourages the integration of knowledge, ideas and technologies from both inside and outside the company. The aim is to take advantage of external resources, such as start-ups, universities, suppliers and even customers, to speed up the innovation process and increase its diversity.

III.2.2. Features :

• **External collaboration:** Companies collaborate with external partners, such as research institutes, start-ups or customers, to co-create new ideas or technologies.

• **Risk sharing:** The costs and risks associated with innovation can be shared between several partners.

• **Access to diversified resources:** By opening up to the outside world, companies gain access to resources, skills and technologies that they do not have in-house.

• **Flexibility:** Open innovation means that technologies or ideas developed elsewhere can be incorporated quickly, reducing the time it takes to bring an innovation to market.

III.2.3. Advantages :

• **Accelerating innovation:** By exploiting external ideas and technologies, companies can accelerate the development and marketing of new products or services.

• **Cost reduction**: Sharing development costs with external partners can reduce the financial burden for everyone involved.

• **Improved diversity of ideas:** By integrating external perspectives, innovation becomes richer and more diverse, increasing the chances of success.

• **Adaptability:** Companies can more easily adapt to market changes by adopting external innovations.

III.2.4. Disadvantages:

• **Confidentiality issues:** Sharing information with external partners may entail risks of disclosure or theft of intellectual property.

• **Complexity of management:** Coordination between several external partners can become complex and require major efforts in terms of project management and communication.

• **Benefit sharing:** The benefits of innovations often have to be shared with partners, which can reduce the gains for each party.

Table 1.1: Comparison between Closed and Open Innovation.

Aspect	Closed Innovation	Open Innovation
Innovation process	Internal and exclusive	Internal and external collaboration
Control	Total by the company	Shared with partners
Idea Security	Very high	Less secure, risk of leakage
Access to Resources	Limited to internal resources	Access to a wide range of external resources
Development time	Longer	Potentially shorter
Costs	Higher	Shared with partners
Diversity of Ideas	Limited to internal ideas	Greater diversity thanks to external input

CHAPTER 2
THE LEGAL REGIME OF PATENTS

I. Legal protection of patents :

At the heart of the drafting and filing process are the key principles that guarantee the legal protection of intellectual property in the form of a patent. Together, these principles ensure that only innovations that meet strict criteria will benefit from adequate protection.

- **What's new :**

This principle requires that the invention must not have been made public before the patent application was filed. No information about the invention must be accessible to the general public via publications, conferences, sales or demonstrations.

- **Inventive step :**

The question is whether the invention represents a non-obvious advance for a person skilled in the technical field concerned. It is essential that the invention constitutes an advance over the existing state of the art.

- **Industrial applicability :**

An innovation must be usable or produced in an industrial field. It must have a practical use and a real application.

- **Detailed description of the invention :**

The patent application must contain a complete and precise description, with well-defined technical terms, setting out the interest of the invention. This enables experts in the field to understand it unambiguously.

- **Specific demands:**

The claims define the limits of the protection granted by the patent. They must be clear, succinct and specific in order to describe the very essence of the invention.

- **Public disclosure and publication :**

Once granted, the patent is generally made public, disclosing the technical details of the invention. This ensures transparency while offering legal protection.

• **Limited period of protection :**

Patents offer protection for a defined period, often 20 years from the date of filing of the patent application. When this period expires, the invention enters the public domain, accessible to all for free use.

II. Patent holders :

Patent holders hold exclusive rights over their intellectual property, which entails specific privileges and important responsibilities relating to the protection, exploitation and disclosure of their inventions. Their status is accompanied by both the advantages and responsibilities inherent in this position.

II.1. Rights and responsibilities of patent holders :

• **Exclusive right of use :**

As patent owners, inventors enjoy the exclusive right to exploit their invention. This gives them the power to prohibit others from making, using, selling or importing their invention without authorisation. This exclusive right gives them a competitive advantage, ensuring that their invention can be marketed safely, protected against copying and unauthorised use.

• **Duty of disclosure :**

Patent owners are required to disclose the technical details of their invention in the patent application. This requirement includes a full and accurate description of the invention, as well as claims delimiting the protection sought. The purpose of this disclosure is to promote the dissemination of technical knowledge and to enable other players in the field to build on these advances.

• **Maintaining intellectual property rights :**

Patent holders have a responsibility to maintain their intellectual property rights by paying the required maintenance fees and complying with patent renewal procedures. Keeping an eye on the patent's validity period and ensuring that it remains valid means that exclusive rights can be enjoyed for the period intended."

II.2. Advantages and responsibilities of being a patent holder :

II.2.1. Advantages :

• Legal protection :

The legal protection afforded to patent holders is a major advantage. It provides a legal barrier against unauthorised copying and exploitation of their inventions, giving them temporary exclusivity to market their innovation. This protection also prevents third parties from benefiting from work carried out without prior consent.

• Recognition and credibility :

The status of patent holder brings official recognition of innovation and contribution to technological development. This increased recognition can not only enhance the inventor's credibility in his field, but also improve his reputation with the professional community and the public.

• Revenue potential :

Patents offer lucrative potential, whether through direct exploitation of the invention or through licences granted to other companies. The royalties received through these licences can represent a significant source of income for patent holders, contributing to the profitability of innovation.

II.2.2. Responsibilities :

• Disclosure of technical details :

Patent holders are responsible for fully disclosing the technical details of their invention in the patent application. This disclosure promotes the dissemination of knowledge and enables others in the field to build on these earlier advances.

• Respect for the rights of others :

Respecting the intellectual property rights of others is another major responsibility for patent holders. They must carry out thorough research to avoid infringing other existing patents, and so respect the rights of others in their innovation.

III. Intellectual property rights associated with patents :

III.1. Protection mechanisms :

Patents are a form of legal protection granted to owners, giving them exclusive rights over their inventions. They encompass fundamental mechanisms for securing the intellectual property rights associated with inventions:

- **Exclusive right of use :**

Patent holders have an exclusive right to exploit their invention and restrict any unauthorised use by third parties. This exclusivity allows them to control the marketing and exploitation of their creation, offering a significant competitive advantage.

- **Protection against counterfeiting :**

Patents provide a defence against infringement, preventing unauthorised use of the invention by third parties. Patent holders are entitled to take legal action against any person or company infringing their intellectual property rights. The courts can award damages to patent holders in cases of proven infringement.

- **Transfer of intellectual property rights :**

The intellectual property rights attached to patents may be assigned or licensed to other entities. Patent owners can sell or transfer their patents to third parties, allowing them to exploit the invention in exchange for royalties or other financial compensation. They may also grant licences to third-party companies to use the invention in return for royalties.

III.2. Period of protection :

The period of patent protection is limited and requires renewal procedures, the two main aspects of which are as follows:

- **Period of protection :**

A patent is usually protected for around 20 years from the date on which the application is filed. During this period, the patent holder has exclusive rights to exploit his invention and enjoy his intellectual property rights.

- **Renewal obligations :**

Maintaining patent protection involves paying maintenance fees at regular intervals. These fees, paid annually or at specific intervals, vary according to the country and the period of validity of the patent. Failure to pay these fees may result in the expiry of the patent and the loss of its exclusive privileges.

Patent owners need to be vigilant when it comes to renewal dates and payment of maintenance fees. Patent offices usually send payment reminders, but it is the patent holder's responsibility to meet the deadlines and make the required payments on time.

III.3. Legal aspects of patents under Moroccan law :

Law 17.97, amended and supplemented by Law 31.05, constitutes the fundamental legal framework for the management of patents and industrial property in Morocco. It defines filing procedures, the term of protection, the rights and obligations of owners, and the procedures for contesting and litigating patents. The recent amendments to this legislation respond to the challenges posed by technological developments and the need to protect inventors, while adapting the criteria for patentability and strengthening the rights of holders.

III.3.1. Overview of the specific legislation in force :

Law 17.97 on the protection of industrial property, amended and supplemented by Law 31.05, constitutes the main legal framework governing patents and industrial property in Morocco. The aim of this legislation is to provide adequate protection for inventors, while encouraging technological innovation. Here is an overview of the key provisions of this legislation concerning patents:

• Patent filing procedure :

The legislation lays down clear rules for filing patent applications, including form and content requirements. It specifies the filing procedures and associated fees. It also defines the criteria for patentability, such as novelty, inventive step and industrial application.

• Duration of Patent Protection :

The law specifies that the term of protection for patents is generally 20 years from the date of filing of the application. There are also periodic renewal requirements, with fees payable to maintain the validity of the patent.

• Content and Limits of Protection :

The legislation delimits the scope of protection conferred by a patent, specifying the elements that can be claimed and the conditions for extended protection. It also includes specific grounds for rejecting patent applications, particularly in areas such as software, medicines and business methods.

• Rights and Obligations of Patent Owners :

Patent holders enjoy exclusive rights, including the right to prohibit unauthorised exploitation of their invention by third parties. The legislation also imposes obligations, such as disclosing the technical details of the invention and respecting the intellectual property rights of others.

• Dispute and litigation procedures :

Specific procedures for challenging the validity of a patent are provided for, including oppositions or administrative appeals. The law also establishes judicial procedures for enforcing intellectual property rights and resolving patent disputes.

III.3.2. Amendments and additions to take account of technological developments and the protection needs of inventors :

Law 17.97 as amended and supplemented by Law 31.05 (or any other relevant legislation) may include amendments and supplements that reflect technological developments and the protection needs of inventors. Such amendments may include:

• Extension of patentability to new areas of technology :

The legislation has been adapted to include advances in emerging fields such as information technology, biotechnology and artificial intelligence, thereby broadening the scope for patentability.

• Adaptation of patentability criteria :

Adjustments have been made to the patentability criteria to take account of the specific features of certain areas of technology, such as the inventiveness of business methods or the protection of scientific discoveries.

• Strengthening the rights of patent holders :

Legislation has been amended to introduce additional safeguards against counterfeiting and to increase penalties for infringement of intellectual property rights.

• **Simplification of patent filing and renewal procedures :**

Changes have been made to simplify the procedures for filing and renewing patents, in particular by introducing online tools and accelerated mechanisms for processing applications.

III.4. **International legislative framework for patents :**

The international legislative framework relating to patents is mainly governed by a number of international treaties and conventions which aim to harmonise and standardise patent protection systems throughout the world. Among the most important are :

• **The Paris Convention for the Protection of Industrial Property (1883) :**

The Paris Convention is one of the first international treaties on intellectual property. It establishes fundamental principles such as the right of priority, which allows a person who has filed a patent application in a member country to claim that filing date in other member countries within a 12-month period. This treaty is essential for inventors wishing to protect their inventions internationally.

• **The 1994 Agreement on Trade-Related Aspects of Intellectual Property Rights (TRIPS):**

The TRIPS Agreement, which forms part of the World Trade Organisation (WTO) agreements, imposes on WTO members minimum standards for the protection of intellectual property rights, including patents. It requires that patents be available for any invention, whether a product or a process, in all fields of technology, and that protection be granted for a minimum period of 20 years from the date of filing.

• **The 1970 Patent Cooperation Treaty (PCT):**

The PCT allows inventors to file a single international patent application that has the same effect as a national application in several member countries. This treaty facilitates the procedure for inventors by giving them more time to assess the commercial potential of their invention before having to incur significant expenses for patent filings in several countries.

• **The Madrid Protocol Concerning the International Registration of Marks (1989) :**

Although primarily concerned with trade marks, the Madrid Protocol has important implications for patents in that it provides a structure for the protection of trade marks in several jurisdictions on the basis of a single application, thereby simplifying the protection of inventions associated with trade marks.

III.5. Legislative framework for patents in France :

In France, the legislative framework relating to patents is mainly governed by the Intellectual Property Code, which incorporates the obligations arising from international treaties, as well as European legislation.

• The Intellectual Property Code (CPI):

The CPI governs all intellectual property rights in France, including patents. It defines the conditions under which inventions can be patented, the filing and examination procedures, and the rights conferred by a patent. In France, an invention can be patented if it is new, involves an inventive step and is capable of industrial application.

• National Institute of Industrial Property (INPI):

INPI is the French body responsible for managing patents. It receives patent applications, examines them and grants patents. The INPI also plays a key role in raising awareness and providing training in industrial property issues.

• The system of utility certificates :

In addition to patents, the CPI provides for a system of utility certificates, which offer 10-year protection for inventions without the need for in-depth examination. This system is often used for inventions with a shorter life cycle.

• Patent case law and litigation :

France has a wealth of case law on patents, with specialised courts dealing with intellectual property disputes. Patent infringement actions are frequent and can result in injunctions, damages and the invalidation of patents.

• European influences :

France is a member of the European Patent Office (EPO), and European patents granted by the EPO have the same effect as a national patent in France. France also participates in the European unitary patent system, which aims to provide uniform protection in several European countries.

- **Recent developments :**

France has recently reformed its patent law to comply with European directives and to strengthen the protection of inventors' rights. These reforms include the introduction of post-grant opposition, allowing third parties to challenge the validity of a patent after it has been granted by the INPI.

CHAPTER 3.
PATENT REGISTRATION

I. Protecting inventions by filing patents :

Filing a patent application is a fundamental step in securing the rights to new inventions and discoveries. To guarantee the validity and effectiveness of your protection, it is crucial to follow a structured and methodical process. The first step in this process is to gather and analyse the relevant information. This requires a thorough search to identify any similar inventions and to confirm the novelty of your invention. Once this information has been gathered, the next step is to draft the patent application, which must be clear, precise and exhaustive.Although the criteria for patentability vary from country to country, three basic conditions must generally be met for an invention to be considered patentable These are novelty, inventive step and industrial application. These criteria form the basis on which patents are granted, ensuring that only truly innovative inventions obtain this legal protection.

II. Patent filing procedures in Morocco :

II.1. Key stages and procedures :

II.1.1. How and where to file a patent application :

In Morocco, a patent application can be filed either directly at the headquarters of the Office Marocain de la Propriété Industrielle et Commerciale (OMPIC) in Casablanca, or via the online patent filing platform. Online filing requires the prior installation of a plug-in, as well as the use of a personal smart card to guarantee the security and integrity of the application. When filing electronically, in accordance with Article 2 of the Decree, it is not necessary to submit a paper version of the application. OMPIC's online platform is accessible around the clock, allowing users to file their applications at any time, 24 hours a day, including public holidays. As soon as the application is submitted, OMPIC immediately provides an acknowledgement of receipt, specifying in particular the application number, the nature of the documents submitted and the date and time of receipt. The date of registration of the documents on the platform is considered to be the official filing date.

II.1.2. Eligibility to file a patent application :

Any natural or legal person, whether resident in Morocco or not, is eligible to file a patent application. An application may be filed in the name of one or more persons, in Arabic or French. It is imperative that all documents submitted are in the same language. The right to a patent shall vest in the inventor or his successors in title, subject to the provisions of Article 18. In the event of the invention being made independently by several persons, the right shall be granted to the person who proves the earliest filing date, in accordance with Article 16. Entry of the application in the National Patent Register shall confer on the applicant all the rights relating to ownership of the patent. However, if the application has been made in violation of the rights of the inventor or his successors in title, or in breach of a legal or contractual obligation (as in the case of employees' inventions governed by Article 18), the applicant may be deprived of his right to the patent, in particular in the event of a claim of ownership in accordance with Article 19.

II.1.3. Preparing the application :

According to Article 31, a patent application file must include the following elements:

- **B1 filing form:** Downloadable via the link provided by OMPIC.
- **Description of the invention :** Technical details of the invention.
- **Claims :** Definition of the aspects of the invention for which protection is sought.
- **Drawings:** Illustrations associated with the description or claims, if necessary.
- **Abstract:** Brief summary of the invention.

The initial description may be submitted in any language at the time of filing, but must be regularised in Arabic or French by the deadline.

II.1.4. Examination of the application :

The official filing date is only given if the documents submitted meet the following minimum conditions:

- **Compliant application form:** duly completed in accordance with article 4 of the decree implementing law 17-97.
- **Description of the invention:** Either a full description or a reference to an accessible earlier application.

If these minimum conditions are not met, the application will be deemed inadmissible (article 31). In the case of electronic filing, if the application is submitted on a public holiday or non-business day, the official filing date will be that of the next business day, subject to the conformity of the documents submitted.

II.1.5. **Regularisation of the application :**

If the application is incomplete, the applicant or his representative has three months from the date of filing to put it right (article 32). In the event of filing with priority claim, the supporting document must be submitted within four months of the expiry of the earliest priority period (Article 8). A patent application which has not been regularised within the time limits is considered to have been withdrawn. Other documents such as the claims, the abstract and the representative's mandate, if appointed, may be submitted within the following deadlines:

- On the initial filing date.
- Within three months of the filing date.
- Within five months of the date of filing, subject to the submission of a motion to continue proceedings.

II.2. **Patentability criteria :**

The patentability criteria are key elements to be taken into account when filing a patent application. They determine whether an invention is eligible for the legal protection conferred by a patent. Although the criteria may vary slightly from one country to another, they are generally based on the following principles:

- **New product :**

For an invention to be considered patentable, it must be new. This means that the invention must not have been made available to the public before the filing date of the patent application. Novelty means that the invention must not have been disclosed, published or presented to the public in any form whatsoever (publications, presentations, sales, etc.). Any prior disclosure by the inventor himself may also compromise the novelty of the invention. This criterion is essential to ensure that patent protection is only granted to truly novel innovations.

- **Inventive activity :**

Inventive step is a criterion that assesses whether the invention presents a non-obvious advance compared with the existing state of the art. To satisfy this criterion, the invention must not be a simple combination or obvious modification of techniques or knowledge already known in the technical field concerned. A person of average skill in the field should not be able to reproduce or deduce the invention easily from existing knowledge. This criterion ensures that patents are only granted for inventions that make a genuine technical contribution, beyond that which is obvious to a specialist in the field.

- **Industrial applicability :**

To be patentable, an invention must also be industrially applicable. This means that the invention must be capable of being made or used in an industrial or commercial field. Industrial applicability ensures that the invention is sufficiently concrete to be implemented in a practical setting, rather than remaining a theoretical or abstract idea. This criterion aims to promote innovations that can be exploited in a tangible way, thus contributing to industrial and economic development.

II.3. Drafting and structure of a patent application :

II.3.1. Drafting a patent application :

Drafting a patent application is an important step that must be carried out with the applicant's specific objectives and overall filing strategy in mind. A well-drafted patent application not only ensures that the invention is protected, but also maximises the chances of success when the application is examined.

A typical patent application includes the following elements:

- **A detailed description of the invention**: This section should explain in depth the operation, structure and technical advantages of the invention.
- **Claims**: These define the limits of the legal protection sought for the invention.
- **Technical drawings**: These illustrate the various configurations or embodiments of the invention, making it easier to understand the technical aspects.

Before starting the actual drafting, it is essential to carry out an in-depth analysis of the invention's patentability. This step enables you to check that the invention meets the criteria for patent protection, and to guide the drafting strategy accordingly.

The following steps are essential for drafting a patent application:

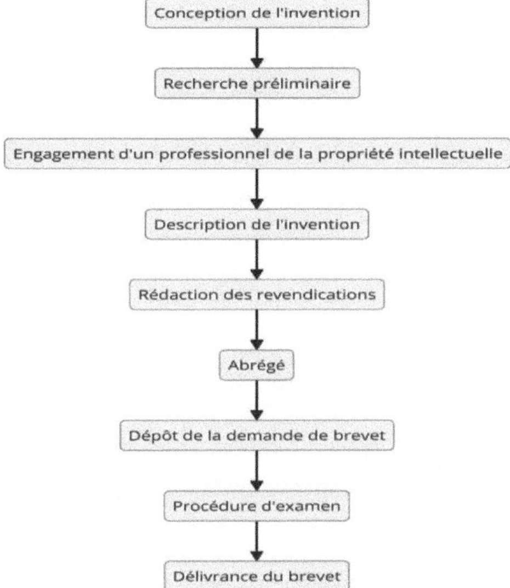

Figure 3.1. Stages in the patent application process.

i. **Conception of the invention** :

Before drafting, it is essential to understand the distinctive aspects of the invention. This includes identifying the technical and functional features that make it unique in relation to the state of the art.

ii. **Preliminary research :**

A preliminary search should be carried out to ensure that the invention is new and not obvious. This search may involve consulting patent databases and other technical sources to identify any relevant prior disclosures.

iii. **Hiring an intellectual property professional :**

It is strongly recommended that you call on the services of a patent lawyer or patent agent to draft the application. These professionals have the expertise needed to draft an application that complies with legal requirements and maximises the chances of obtaining a patent.

***iv.* Description of the invention:**

The description must be clear, complete and precise, detailing the operation, structure, components and advantages of the invention. The use of drawings or technical diagrams is often useful to facilitate understanding.

***v.* Drafting of claims :**

The claims are the most crucial part of the application, as they define the aspects of the invention that will be protected by the patent. They must be drafted in such a way as to cover the unique features of the invention, while remaining broad enough to include different possible variants.

***vi.* Abbreviated :**

The abstract provides a succinct overview of the invention and its advantages. It should be concise and provide a clear idea of the invention without going into technical detail. The abstract is often used in subsequent searches to identify relevant patents.

***vii.* Filing of the patent application :**

Once the drafting is complete, the patent application must be filed with the competent body, either a national or regional patent office, depending on the applicable system.

***viii.* Examination procedure :**

After filing, the application undergoes an examination procedure by the competent body. This phase may include exchanges with the examiner and amendments to the application to meet the patentability criteria.

***ix.* Issue of the patent :**

If the application is judged to meet the criteria, a patent is granted, giving the holder the exclusive right to exploit the invention for a period of generally 20 years from the date of filing. It is important to note that a poorly drafted patent application can lead to difficulties in enforcing the associated rights or in effectively protecting the invention. Drafting a patent application is therefore a complex task that can vary from one country to another, often requiring the intervention of qualified professionals.

II.3.2. Structure of a patent application :

The structure of a patent application is generally standardised and comprises the elements shown in Figure 3.2.

Figure 3.2. Structure of a patent application.

i. Title page :

This page includes the title of the invention, the names of the applicant and inventors, as well as information on the priority claimed, where applicable.

ii. Abbreviated :

The abstract is a brief summary of the invention, highlighting its main features and advantages. It provides a quick overview of the invention.

iii. Description of the invention and its technical field :

This section details the operation, structure and technical characteristics of the invention. It may be accompanied by drawings, diagrams or photographs for better understanding.

iv. Claims :

The claims specify the technical characteristics for which protection is sought. They are crucial in defining the scope of the legal protection granted by the patent.

v. Drawings or diagrams :

Although optional, drawings or diagrams may be included to illustrate and complete the description of the invention. They must be numbered and accompanied by explanatory captions.

***vi.* Abbreviations and glossary :**

If abbreviations or specific technical terms are used, it may be useful to include a glossary to explain their meaning.

***vii.* References cited :**

This section lists the relevant reference documents, such as earlier patents or publications, used during the preliminary search or relevant to understanding the invention.

***viii.* Nucleotide or amino acid sequences :**

If the invention concerns genetic sequences or biological molecules, these must be included in a separate section.

***ix.* Priority information :**

If the application claims priority over an earlier application, the relevant information must be included in a dedicated section.

***x.* Information on the inventors and the applicant :**

This section must include the names and addresses of the inventors, as well as the contact details of the applicant.

III. International patent filing process :

The international patent filing process is structured to allow inventors to protect their inventions in several countries simultaneously, while respecting the local requirements of each jurisdiction. The main steps in this process include the use of systems such as the Patent Cooperation Treaty (PCT) and the Paris Convention.

The general steps involved in filing a patent internationally :

Step 1. Prior art search :

• **Objective**: before filing a patent application, it is crucial to carry out a worldwide prior art search to ensure that the invention is new and has not already been patented elsewhere.

• **Tools**: use of international databases such as those of the World Intellectual Property Organisation (WIPO), the European Patent Office (EPO) and others.

Step 2. Filing of the first patent application (priority) :

• **Place**: the first filing is generally made in the inventor's country of origin. This application serves as a reference for the priority date.

• **Time limit**: under the Paris Convention, once this application has been filed, the inventor has 12 months to file applications in other countries, claiming the priority date of the first filing.

Step 3. Choice of international filing strategy :

• **Option 1: direct national filing**

o **Process**: filing of separate patent applications in each country or region where protection is sought.

o **Complexity**: each application is subject to local laws and requirements, requiring individual translations and fee payments.

• **Option 2: filing via the Patent Cooperation Treaty (PCT)**

o The **aim is** to simplify the filing of patents in several countries via a single procedure.

o **Procedure**: filing a pct application which allows the invention to be protected in up to 156 member countries, delaying the need to submit specific national applications.

Step 4. Filing the PCT application :

• **Where**: the application can be filed with the WIPO, the European Patent Office (EPO) or the national industrial property office of the inventor's country of origin.

• **Content**: the application must include a full description of the invention, claims, an abstract and, if necessary, technical drawings.

• **International search report**: after filing, WIPO carries out an international search to identify relevant prior art.

Stage 5. International publication :

• **When**: 18 months after the priority date, the PCT application is published by WIPO.

• **Effect**: publication makes the invention accessible to the public and marks the start of the phase of entry into the various national or regional jurisdictions.

Step 6. International preliminary examination (optional) :

• **Objective**: the applicant can request an international preliminary examination to obtain a non-binding opinion on the patentability of the invention.

• **Advantage**: this enables the application to be strengthened before entering the national phases and to better assess the chances of obtaining a patent.

Stage 7. Entry into the national or regional phase :

• **When**: 30 months after the priority date (or 31 months in some countries), the applicant must enter the national or regional phases to pursue protection in the chosen jurisdictions.

• **Procedure**: each national or regional office examines the application according to its own rules. This may include the payment of fees, the provision of translations, and adjustments to the claims.

Stage 8. Examination of the application by the national/regional offices :

• **Examination**: each office carries out a substantive examination to check that the invention meets the patentability criteria (novelty, inventive step, industrial applicability).

• **Responding to objections**: the applicant may have to respond to objections or amend claims to comply with local requirements.

Stage 9. Issue of patents :

• **Decision**: if the examination is conclusive, the national or regional offices grant a patent.

• **Duration**: patents are generally valid for 20 years from the date of filing, subject to payment of annual or maintenance fees.

Step 10. Payment of annual taxes :

• **Requirement**: the patent holder must pay annual fees in each country where the patent is maintained in order to maintain the validity of the protection.

• **Consequence**: non-payment of fees leads to the forfeiture of patent rights in the jurisdictions concerned.

Stage 11. Monitoring and advocacy :

• **Monitoring**: it is essential to monitor the exploitation of the invention to detect any infringement.

• **Litigation**: in the event of infringement of rights, the holder may take legal action in the countries where the patents were granted.

Filing a patent internationally is a complex and costly process, but essential for inventors and companies wishing to protect their innovations in several countries. A well-planned strategy, including the use of the PCT, can simplify this process and offer effective protection on a global scale.

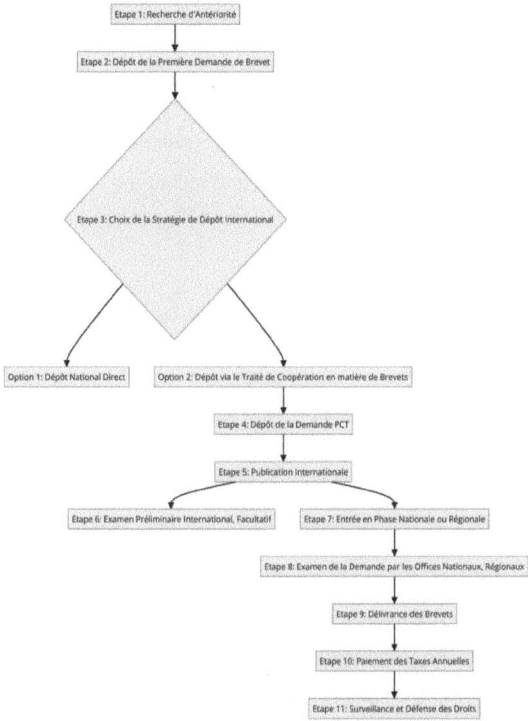

Figure 3.1. The international patent filing process.

CHAPTER 4
STRATEGIC PATENT MANAGEMENT :DEVELOPMENT, LICENSING AND PROTECTION

I. Making the most of patented inventions :

The technological valorisation of patents is a key process for maximising the scientific, technical, economic and strategic value of an invention. This process relies not only on technology assessment and development, but also on the integration of technology intelligence and technology transfer to transform innovations into marketable solutions. In this chapter, we explore strategies for the technological valorisation of patents, highlighting the importance of technology watch and technology transfer in this context.

I.1. Technology watch :

Technology watch is a basic tool in the patent development process. It involves monitoring developments in technologies, markets and the activities of competitors. By conducting an effective technology watch, companies can identify emerging trends, market opportunities and potential threats. This information can be used to guide strategic decisions relating to patent development. Technology watch plays several roles in the development of patents:

- **Identifying opportunities**: By monitoring market trends, companies can identify untapped niches where their patented technology could be exploited.
- **Adapting our strategy**: Rapidly evolving technologies require us to continually adapt our development strategies. Monitoring enables us to adjust our development plans in line with market developments.
- **Risk prevention**: By anticipating competing innovations, companies can protect their market position and prevent their technology from becoming obsolete.

I.2. Technology transfer :

Technology transfer is the set of activities aimed at transforming an invention into a commercially viable product or service. This process involves several stages, from research and development to the protection of intellectual property and marketing.

The main strategies for exploiting patents include :

• **Technological Maturation**: This involves adding value to an invention through further technical development, the creation of prototypes and the international protection of intellectual property rights. This phase aims to demonstrate the technical and commercial feasibility of the technology.

• **Licensing**: Licensing enables a patent to be exploited commercially by granting other companies the right to use the technology. The terms of the Licensing agreements (exclusivity, duration, territory) are essential to maximise revenues and manage the risks associated with exploiting the technology.

• **Company creation**: The creation of start-ups or spin-offs is a common strategy for adding value to R&D results. These new companies can focus on developing and marketing patented technologies, while benefiting from the technical expertise and resources of the original research organisation.

• **Collaboration and Partnerships**: Collaborations with other companies or research institutions can facilitate technological development and access to markets. Joint ventures and strategic partnerships are examples of such collaborations, enabling resources and skills to be combined for mutually beneficial added value.

I.3. Technology transfer :

Technology transfer is a fundamental element in the exploitation of patents. It refers to the process of transferring knowledge and technologies developed in a research context to industry, where they can be commercially exploited.

The various forms of technology transfer include :

• **Technology sale or licence**: The patented technology can be sold or licensed to another company capable of marketing it effectively. This enables the invention to be monetised quickly while limiting the risks for the patent holder.

• **Research and Development (R&D) agreements**: These allow companies to collaborate with research institutions to further develop a technology and prepare it for commercialisation. In return, the companies obtain intellectual property rights or share in the revenues generated by the exploitation of the technology.

• **Franchising and Joint-Ventures**: These technology transfer strategies enable structures to be put in place to exploit a technology commercially on a large scale, by bringing together industrial partners who contribute complementary resources.

• **Incubators and Accelerators**: These structures offer support to innovative start-ups, facilitating the development of new technologies and their introduction to the market. They play a crucial role in technological maturation and networking with industrial partners.

II. Technology roadmap :

Adding technological value to a patented invention is a complex process that requires strategic planning and methodical implementation. The technology roadmap serves as a guide to maximising the commercial potential of an innovation, by defining key stages, milestones and development strategies. This chapter presents optimised roadmap structuring, including strategic planning, technology assessment, definition of objectives, development of the commercialisation strategy, and management of the development stages.

II.1. Strategic Planning for Technology Transfer

II.1.1. Technology assessment :

In-depth technology assessment is the first stage in strategic planning. It involves analysing the technical characteristics, performance, durability and competitiveness of the invention compared with existing solutions on the market. Laboratory tests, simulations and feasibility studies are used to identify the strengths and weaknesses of the technology. This assessment also includes an economic analysis, taking into account production costs and potential economies of scale. The aim is to provide a solid basis for defining value-adding objectives and development strategies.

II.1.2. Definition of recovery objectives :

The exploitation objectives are the concrete results that we hope to achieve when the technology is commercialised. They must be clear, specific, measurable, achievable, relevant and time-defined (SMART). Common objectives include

• **Market expansion**: Introducing the technology to new geographical markets or new sectors.

• **Acquiring market share**: Becoming a major player by capturing a significant share of the market.

• **Improving profitability**: increasing profit margins and reducing production costs.

• **Protecting intellectual property**: Strengthening intellectual property rights to preserve competitive advantage.

• **Strategic partnerships**: Working with other companies to accelerate time to market.

• **Creation of a stand-alone company**: Establish a new entity dedicated to exploiting the technology.

II.2. Developing a value-adding strategy :

II.2.1. Market analysis :

Market analysis is important for understanding the trends, needs and competitive dynamics of the sector. This analysis helps to identify the most promising market segments, assess the demand for the invention, and determine the best approaches for penetrating the market. It serves as a basis for developing a positioning strategy that differentiates the technology from competitors.

II.2.2. Choice of business models :

The choice of business model needs to be aligned with the value creation objectives. There are several possible options:

• **Direct sales**: Direct marketing to consumers or businesses.

• **Licensing**: Granting licences to third parties to exploit technology in exchange for royalties.

• **Strategic partnerships**: Working with other companies to share resources and accelerate development.

• **Creation of a start-up**: Formation o f a new company to exploit the technology independently.

II.2.3. Development planning :

A detailed development plan must be drawn up, including the design of prototypes, validation tests and industrialisation. This plan must define the critical stages of development, the resources required, and the deadlines to be met in order to achieve the valorisation objectives.

II.2.4. Communication and marketing :

The communications strategy must be designed to effectively promote the technology in the marketplace. This includes developing marketing materials, participating in trade shows, and managing public relations to generate interest and establish a solid reputation.

II.2.5. Risk management :

A proactive assessment of valuation risks is essential. Risks can include competition, regulatory constraints, technological challenges and financial barriers. Mitigation plans must be put in place to manage these risks effectively.

II.3. Identification of key stages and development milestones :

II.3.1. Technical development stages :

Technical development is structured in several stages:

- **Research and preliminary analysis**: Assessment of technical feasibility.
- **Design and prototyping** : Development of prototypes for test functionality.
- **Tests and validation**: Validation of performance and compliance with standards.
- **Industrialisation**: Optimisation of manufacturing processes for mass production.
- **Continuous improvement**: Post-market adaptation and improvement.

II.3.2. Stages of business development :

Business development includes :

- **Market research**: in-depth analysis to understand the target market.
- **Marketing strategy**: Development of a marketing plan to position the technology.
- **Sales plan** : Developing an effective sales strategy.
- **Market launch**: Launch activities to introduce the technology.
- **Partnership development**: Establishing strategic collaborations.
- **Monitoring and adaptation**: Continuous monitoring of performance and any necessary

adjustments.

II.3.3. Development milestones :

Development milestones are critical points that mark the progress of the project:

• **Validation of technical feasibility**: Confirmation of technical capabilities.

• **Obtaining a patent**: legal protection of the invention.

• **Initial financing**: Securing the necessary funds.

• **Functional prototype**: Development and validation of the prototype.

• **Validation tests**: Tests under real conditions.

• **Market launch**: Introduction to the market.

• **Expansion**: Phase of growth and penetration of new markets.

III. Strategies for exploiting patents :

Patents are a fundamental tool for protecting inventions and giving companies a competitive edge. However, for a patent to become a real added value, it is essential to put in place effective exploitation strategies. Adding value to patents is not limited to simply protecting them legally, but also involves exploiting the invention economically to maximise revenue and strengthen the company's market position. Before putting in place a strategy for adding value to patents, it is important to carry out an in-depth analysis of the patents held by the company. This analysis must take into account :

• **Market potential**: This involves assessing current and future demand for the patented invention. The valuation of a patent will be more effective if the target market is expanding or if the invention meets an unmet need.

• **Competitive positioning**: Analysing the position of the patent in relation to the competition is essential. A patent that covers a unique technology or product, which is difficult for competitors to circumvent, has a higher potential value.

• **The life of the patent**: the closer a patent is to expiry, the less likely it is to be valued effectively. The analysis must therefore incorporate strategies for maximising revenues before the end of the protection period.

• **Geographical coverage**: The protection offered by a patent is limited to the jurisdictions in which it is filed. Successful exploitation requires geographical coverage in line with the target markets.

III.1. Partnerships and strategic alliances :

One of the most common strategies for enhancing the value of a patent is to form partnerships or alliances with other companies. These partnerships can take several forms:

• **Patent licences**: By granting a licence, the patent holder allows another company to exploit the invention in exchange for royalties or a lump-sum payment. This strategy makes it possible to generate income while retaining the rights to the invention. It is essential to choose partners carefully and negotiate the terms of the licence to maximise financial returns.

• **Co-development and co-marketing**: In this approach, two companies join forces to jointly develop and market a product or technology based on a patent. This makes it possible to share costs and risks while benefiting from the resources and expertise of both partners.

• **Creation of joint ventures**: Joint ventures enable companies to combine their resources to exploit a patent more effectively. This approach is particularly useful for penetrating new markets or developing complex technologies requiring significant investment.

III.2. Cross-licensing and patent portfolios :

Cross-licensing involves an exchange of licences between two or more companies, allowing them to use each other's patents. This strategy is particularly effective in industries where technologies are complex and interdependent. Cross-licensing can reduce litigation costs, encourage innovation and provide access to new technologies.

Managing a portfolio of patents, rather than an individual patent, is also a powerful strategy for adding value. A well-managed patent portfolio can :

• **Strengthen negotiating position**: A diversified portfolio offers greater flexibility when negotiating licences or partnerships, increasing the chances of obtaining advantageous terms.

• **Maximise revenues**: By offering licences for a complete portfolio, rather than for individual patents, a company can increase the perceived value and thus justify higher royalties.

• **Reduce risks**: A portfolio of patents provides better protection against litigation, because it is more difficult for a competitor to circumvent several patents than it is to circumvent a single one.

III.3. Marketing strategies :

Direct marketing is another key strategy for adding value to a patent. This involves incorporating the patented invention into the company's products or services and bringing them to market. To succeed with this strategy, a number of factors need to be taken into account:

• **Product development**: The patented invention must be transformed into a marketable product. This may require investment in R&D, prototyping and testing.

• **Marketing positioning**: The product must be positioned to maximise its appeal to consumers. This includes pricing, distribution strategy and promotion.

• **Market protection**: To maximise value, it is crucial to protect the market for the patented invention against competition. This can be achieved by obtaining additional patents, continuously improving the product or strengthening the brand.

III.4. Continuous innovation and technology watch:

Finally, patent valuation should not be a static strategy. Continuous innovation and technological intelligence are essential to maintaining and increasing the value of patents. This means :

• **Continuous improvement of the invention**: By developing improvements or variations on the initial invention, a company can extend the commercial life of the patent and maintain its competitive edge.

• **Monitoring market trends**: An active technology watch enables us to detect market trends and anticipate future needs. This helps us to adapt our development strategy accordingly.

• **Proactive patent portfolio management**: Dynamic management of a portfolio, including the acquisition of new patents and the abandonment of obsolete ones, maximises revenues while minimising costs.

IV. Strategic collaboration in patent exploitation: partnerships, alliances and innovative models :

Collaboration plays an essential role in the exploitation of patents, enabling patent holders to maximise their commercial potential. Partnerships and strategic alliances enable the players involved, whether companies, research institutes or investors, to join forces to exploit and market patented inventions effectively. These collaborations facilitate the sharing of resources, access to complementary skills and optimisation of the value of patents, while stimulating innovation and promoting economic development.

IV.1. Definition of concepts :

• **Collaboration:** In this context, collaboration involves cooperation between different players (patent holders, researchers, companies, investors) to maximise the economic value and commercial impact of patents, via licensing agreements, public-private partnerships, etc.

• **Partnership:** A partnership in the context of patent exploitation is a collaboration between two or more entities, such as companies, research institutions or public bodies, which share resources, knowledge and skills to achieve common objectives in the field of patent exploitation.

• **Co-operation:** Co-operation refers to close collaboration between those involved in the exploitation of patents, with the aim of exploiting and commercialising patented inventions through the sharing of resources, information and joint activities.

• **Strategic Alliance:** This term refers to formal or informal agreements between entities to collaborate on specific aspects of patent exploitation, such as technological development, access to new markets, or the creation of innovative products.

• **Consortium:** A consortium is a form of partnership where several entities come together to collaborate on patent exploitation projects, sharing resources, knowledge and skills to develop innovations and market products.

IV.2. Examples of partnerships and strategic alliances:

• **University-Company Partnership:** A university collaborates with a pharmaceutical company to develop and market a drug based on patented technology.

• **Strategic Automotive Alliance:** A car company teams up with a battery manufacturer to develop innovative electric vehicles.

• **Technology-IA collaboration:** A technology company collaborates with a start-up specialising in artificial intelligence to integrate patented functions into software.

• **Biotechnology Licence Agreement:** A biotechnology company grants a licence to a medical device manufacturer to use patented technology in innovative products.

IV.3. Cooperation between the public and private sectors :

Public-private partnerships (PPPs) are collaborations between the public sector (government, local authorities, public establishments) and private players to carry out projects of general interest, including the exploitation of patents. This cooperation makes it possible to combine the advantages and skills of both sectors to achieve common goals such as technological innovation, economic growth and sustainable development.

• **Access to public resources :**

Government funding: PPPs offer private companies access to government funding to support the development and commercialisation of patented technologies.

Research infrastructure: Private companies can access public research infrastructure, such as laboratories and specialised equipment, to develop and validate patented technologies.

• **Technical and scientific expertise :**

Cutting-edge scientific research: Public institutions provide high-level scientific and technical expertise, boosting patent-related R&D capacity.

Specialist knowledge: Public sector researchers offer in-depth knowledge in specific technical fields, useful for improving and exploiting patented technologies.

• **Benefits for the private sector :**

Access to specific resources: PPPs enable private companies to benefit from public expertise and resources to overcome technical obstacles and develop innovations.

Opportunities for collaboration: Private companies can forge collaborations with public-sector players, boosting their visibility and competitiveness on the market.

Validation and credibility: Working with reputable public institutions gives additional validation to patented technologies, boosting the confidence of investors and commercial partners.

IV.4 Collaborative models for patent exploitation :

Collaborative models for exploiting patents include a variety of strategic approaches that enable patent holders to work with other players to exploit their inventions to the full. These models include licensing agreements, R&D cooperation, the creation of spin-offs, and participation in business networks or consortia.

• Licensing agreements :

Benefits for the patent holder: Revenue generation, geographical expansion and access to the licensee's resources.

Benefits for the licensee: Access to proven technology, reduced costs and risks, and new business opportunities.

• Cooperation in research and development (R&D) :

Sharing resources and skills: Pooling financial, material and technical resources to accelerate technological development.

Mutual benefits: Accelerated development, reduced risk, access to new knowledge and innovations, and profit-sharing.

• Creation of spin-offs and start-ups:

Advantages for the patent holder: Direct control of operations, direct financial benefits and operational flexibility.

The benefits of focusing: specialist expertise, agility in decision-making and access to specific funding.

• Business Networks and Consortiums

Pooling resources and knowledge: sharing costs and risks, as well as technical and commercial knowledge.

Creating synergies: strategic collaboration to amplify business opportunities and strengthen open innovation.

V. Patent protection - Legal framework :

The legal framework for patents is essential for protecting and promoting innovation. It defines the fundamental principles, rights and responsibilities of patent owners, and establishes the associated protection mechanisms. This chapter explores the key aspects of the patent legal regime, covering the fundamental principles, the roles and responsibilities of owners, the protection of intellectual property rights, and legislative developments in response to technological advances.

V.1. Fundamental principles of patent law :

The legal system for patents is based on essential principles that determine the eligibility of an invention for patent protection. These principles ensure that patents are only granted to inventions that represent genuine technological advances and have tangible industrial potential.

V.1.1. Novelty, inventive step and industrial applicability :

• **Novelty:** An invention must be new, i.e. it must not have been disclosed to the public before the date of filing of the patent application. This principle ensures that patents protect novel innovations.

• **Inventive step:** The invention must involve an inventive step that is not obvious to a person skilled in the technical field concerned. This means that the invention must represent a genuine creative contribution in relation to the existing state of the art.

• **Industrial applicability:** The invention must be susceptible of industrial application, which means that it must be capable of being used in an economic sector. This principle ensures that patents are granted for inventions that have the potential for practical and commercial use.

V.1.2. Protecting inventions and stimulating innovation :

The fundamental principles of the patent legal system encourage innovation by offering solid legal protection to inventors:

• **Incentives for research and development:** Patents offer temporary exclusivity, encouraging inventors to invest in R&D.

• **Protection against unauthorised copying and exploitation:** Patents give inventors the right to prohibit unauthorised use of their invention, thereby protecting their efforts and investments.

• **Dissemination of technical knowledge:** Patents require disclosure of the technical details of the invention, which contributes to the dissemination of knowledge and technological progress.

• **Stimulating continuous innovation:** The limited duration of patents encourages inventors to continue innovating in order to maintain their competitive edge.

V.2. Roles and responsibilities of licence holders :

Patent owners play a crucial role as holders of exclusive rights to their inventions. The legal framework gives them important rights, but also imposes significant responsibilities.

V.2.1. Rights of patent holders :

• **Exclusive exploitation rights:** Patent holders have the exclusive right to exploit their invention, which enables them to control its commercialisation and prohibit any unauthorised use by third parties.

• **Duty of disclosure:** In return for this exclusive right, patent holders must disclose the technical details of their invention in the patent application, thereby facilitating the dissemination of knowledge.

• **Maintaining intellectual property rights:** Patent holders must ensure that they pay maintenance fees to maintain their rights to the patent throughout its period of validity.

V.2.2. Benefits and responsibilities of patent holders :

• **Benefits:** Holders benefit from legal protection, official recognition of their innovation, and the potential to generate income through direct exploitation or licensing.

• **Responsibilities:** They must disclose the technical details of their invention, respect the intellectual property rights of others, and maintain their rights in accordance with legal requirements.

V.3. Protection and duration of intellectual property rights :

Patents offer time-limited legal protection, enabling inventors to profit from their inventions while promoting technological progress.

V.3.1. Intellectual Property Rights Protection Mechanisms :

- **Exclusive right of exploitation:** Patents confer a temporary monopoly on the exploitation of the invention, guaranteeing the holders a competitive advantage.
- **Protection against counterfeiting:** Patents offer protection against unauthorised use of the invention, and patent holders can take legal action against counterfeiters.
- **Transfer of intellectual property rights**: Owners can assign or license their patents to others, allowing the invention to be exploited by third parties in exchange for royalties.

V.3.2. Duration of patent protection and renewal requirements :

- **Term of protection:** In general, patents are protected for 20 years from the date of filing. During this period, patent holders have the exclusive right to exploit their invention.

- **Renewal requirements:** Patent holders must pay maintenance fees at regular intervals to maintain the validity of their patent. Failure to pay these fees may result in the loss of their exclusive rights.

V.4. Legislative framework under Law 17.97 (Morocco) :

Law 17.97 on the protection of industrial property, amended by Law 31.05, constitutes the legal framework governing patents in Morocco. This legislation was designed to offer inventors adequate protection and encourage innovation.

V.4.1. Overview of legislation :

- **Patent filing procedure :** Legislation defines the rules for filing patent applications, the criteria for patentability and the filing procedures.
- **Term of protection:** The term of protection is generally set at 20 years, with renewal requirements to maintain patent rights.
- **Rights and obligations of holders:** Holders enjoy exclusive rights, but must also respect certain obligations, such as technical disclosure and respect for the rights of others.

V.4.2. Changes to take account of technological developments :

• **Extension of patentability:** Legislation can be adapted to include new areas of technology such as biotechnology and artificial intelligence.
• **Adaptation of patentability criteria:** Changes can be made to better reflect current innovation needs, particularly in emerging fields.
• **Strengthening rights:** Additional measures can be introduced to protect patent holders against counterfeiting and to facilitate the filing and renewal of patents.

VI. Assessing the economic value of a patent portfolio :

VI.1. Definitions :

• **Patent portfolio:** A patent portfolio represents all the patents held by an entity, whether an individual or a company. Optimised management of this portfolio, combined with a comparative analysis with other companies, makes it possible to determine the economic value of patents and maintain a competitive advantage. It also helps to identify opportunities and threats, such as the expansion of new markets or the emergence of alternative technologies. By monitoring market developments, it becomes possible to adjust strategy to maximise the value of the patent portfolio.

• **Value of a patent:** The value of a patent lies in the potential economic benefit that its exploitation can generate. The holder of a patent can use it to protect its products or grant licences, thereby generating income. The value of a patent varies according to the purpose for which it is valued and the user of the invention. For example, the value of a patent will be different depending on whether it forms part of the assets of a company in difficulty or whether it is valued as part of ongoing production activities. It will not have the same value for a bank, which might be forced to resell it, as for a market player who already has the necessary means of production.

In addition to its financial value, a patent has a non-monetary value. It can, for example, enhance a company's brand image. Obtaining a patent can also increase the inventor's reputation and stimulate innovation.

VI.2. Patent assessment :

In valuing a patent, it is crucial to consider not only the patent as an exclusive right, but also the underlying technology, as well as a company's ability to exploit its complementary assets (i.e. its ability to commercialise the invention). Although it is theoretically desirable to value each patent individually, this can be difficult in practice, especially when dealing with interdependent patents. In many cases, it is preferable to value groups of patents rather than each patent individually.

VI.2.1. Reasons for evaluating patents :

Various internal and external factors may justify the need to determine the value of a patent.

Table 4.1. Internal and external factors in the value of a patent.

Internal factors	External factors
Optimising the costs of a patent portfolio Aligning patent strategy with overall business strategy Remuneration of inventors	Financing Licensing Mergers and acquisitions Accounting (IAS 38) Valuation in enforcement proceedings (insolvency, damages)

VI.2.2. Indicators of the value of a patent :

Research has identified several indicators that can be used to assess the value of a patent. The number of citations, i.e. references to a patent in subsequent documents, is often considered to be a convincing indicator of its value. Analysis of these references reveals a network of links, known as the "network of patent documents". cited patents", used for evaluation purposes. The number of citations of a patent indicates its scientific importance and, therefore, its value.

Other value indicators include :

- The size of the patent family ;
- The life of the patent ;
- The result of oppositions filed against the patent ;
- The number and quality of claims ;
- Area of application;
- Commercial potential ;
- Litigation history.

These factors, or 'indicators of value', can influence the value of a patent positively or negatively, just as features such as location, number of rooms or proximity to schools influence the value of a house.

VI.3. Assessing the value of a patent - Key criteria to consider :

Methods of valuing patents for commercial purposes can be divided into two groups: quantitative methods and qualitative methods. Quantitative methods aim to calculate the monetary value of a patent or patent portfolio and fall into three categories:

• **Cost method:** This method looks at the costs required, whether internal or external, to develop a similar invention and obtain the corresponding patent. It is generally used in accounting.

• **Market method:** These methods value patents by comparing them with prices obtained in recent comparable transactions. They require an active market, comparable exchanges in intellectual property between two independent parties, and sufficient access to information on the prices of these transactions.

• **Income method:** Income methods measure the revenue that a patent could generate. The present value of the patent is calculated on the basis of expected future revenues (less interest).

All three methods have their advantages and disadvantages. It is essential to determine on a case-by-case basis which is the most appropriate. We therefore recommend that you always consult lawyers and experts in the field.

VI.3.1. Qualitative assessment :

Quantitative evaluation provides an estimate of the monetary value of a patent, but it may not be sufficient to formulate strategic recommendations. Qualitative evaluation takes this analysis a step further, highlighting the strengths and weaknesses of the patent and estimating the various factors involved. For example, a quantitative assessment might state that "the patent is worth EUR 50,000", while a qualitative assessment might conclude that "the patent protects a technology". of strategic importance for a promising market; its protection can be implemented effectively, but significant investment is still required". Qualitative assessment methods are mainly used for internal patent management. They are particularly useful for comparing, ranking and prioritising patents within a portfolio or in relation to those of competitors. They can also be used to assess the risks and opportunities associated with patents.

VI.3.1. The various quantitative patent valuation methods :

Table 4.2: Quantitative patent valuation methods

Method	Description
Cost-oriented :	
(1) Historical costs	Sum of the costs required to generate the patent.
(2) Reproduction costs	All the costs necessary for its reproduction.
Market-oriented :	
(3) Market price on an active market	Valuation of patents based exclusively on the amount you would be prepared to pay on an open market.
(4) Analogy methods	Patent value of a known analogue on the market.
(5) Value of econometric	Marginal contribution to the market value of the patent, generally determined econometrically.
Income-oriented :	
(6) Forecast of flows cash flows	Monetary value based on directly calculable cash flows.
(7) Analogy of licence prices	Cash value based on licence payments (analogous).
(8) method of additional benefit	Cash value based on the difference compared to a fictitious company with no patent protection.
(9) method of the residual value	Cash value based on payments after subtraction of pro forma business costs.
Other patent valuation methods :	
Profit-oriented :	
(10) 25% rule	25% of gross margin before or after tax.
(11) Profit sharing	Value in the form of licences as a fraction (%) of profit.
Focused on future value:	
(12) Technological factor	Monetary value to assess the contribution of technology to cash flow.
(13) Real options method	Cash value based on the value of potential commercial options over time.
(14) Decision tree analysis	Monetary value integrating future decisions at predefined stages.
(15) Monte Carlo method	Simulation of the value of a patent based on random draws.

VII. Overview of the Legal Framework for International Patent Protection

International patent protection is governed by a series of treaties, agreements and organisations that facilitate the mutual recognition of intellectual property rights between countries. This legal framework aims to harmonise rules and simplify procedures for inventors wishing to protect their innovations in several jurisdictions.

VII.1. The World Intellectual Property Organisation (WIPO) :

The World Intellectual Property Organization (WIPO) is the United Nations agency responsible for promoting the protection of intellectual property worldwide. It plays a central role in the development and administration of international patent agreements.

VII.1.1. Patent Cooperation Treaty (PCT) :

The Patent Cooperation Treaty (PCT) is one of the main agreements administered by WIPO. It allows inventors to file a single patent application that is valid in more than 150 member countries, simplifying the international patent application process.

• **Unified procedure:** The PCT allows inventors to file a single international application, which is then examined by national or regional patent offices to obtain protection in the designated countries.
• **International search report:** Once the application has been filed, an international search report is drawn up, providing an analysis of the prior art relevant to the invention, which helps the inventor to assess the prospects of a patent being granted in different countries.

VII.1.2. Budapest Treaty :

The Budapest Treaty is an agreement administered by WIPO that deals with the international deposit of micro-organisms for patent purposes. This treaty simplifies the process of depositing these biological materials for inventors seeking international protection.

VII.2. Agreement on Trade-Related Aspects of Intellectual Property Rights (TRIPS):

The TRIPS Agreement, administered by the World Trade Organisation (WTO), establishes minimum standards for the protection of intellectual property rights, including patents, which WTO members must respect.

• **Minimum standards:** The TRIPS Agreement requires WTO members to grant protection to inventions in all fields of technology, provided that they are new, involve an inventive step and are capable of industrial application.
• **Term of protection:** The agreement stipulates that the minimum term of protection for

patents is 20 years from the date of filing of the application.

• **Remedies:** TRIPS obliges members to provide effective legal remedies against patent infringement and to enable patent holders to obtain adequate redress.

VII.3. Regional patent offices :

In addition to international agreements, several regional patent offices facilitate the filing of patent applications in several countries in the region via a unified procedure:

VII.3.1. European Patent Office (EPO) :

The EPO, based in Munich, offers a centralised procedure for granting patents valid in the member states of the European Patent Convention (EPC). Once granted, the European patent must be validated in each member country in order to enter into force.

• **Application procedure:** A patent application can be filed directly with the EPO, where it will be examined to determine whether it meets the patentability criteria.

• **National validation:** Once the European patent has been granted, it must be validated in the designated countries, which may include translating the patent into the official languages of the countries concerned.

VII.3.2. The African Intellectual Property Organisation (OAPI):

OAPI, based in Yaoundé, Cameroon, is a regional organisation that enables the filing and granting of patents valid in its 17 member states, mainly in French-speaking Africa.

N.B. The OAPI offers a uniform system of protection where a patent granted by the OAPI is automatically valid in all Member States without the need for further validation.

VII.3.3. African Regional Intellectual Property Organisation (ARIPO) :

ARIPO, based in Harare, Zimbabwe, manages a regional patent system that allows inventors to file a single application valid in member countries, mainly in English-speaking Africa.

VII.4. **Harmonisation and challenges :**

Despite harmonisation efforts, there are still differences between national and regional patent systems. These differences may relate to patentability criteria, disclosure requirements and appeal procedures. Inventors must

be aware of these variations when seeking to protect their inventions internationally.

VII.4.1. **Challenges :**

- **Cost:** International patent protection can be costly due to filing, translation and maintenance fees in several jurisdictions.
- **Legal complexity**: Differences between legal systems can make the management of patent rights complex, requiring specialist expertise to navigate the various regimes.
- **International litigation:** Patent litigation can become very complex on an international scale, particularly when several jurisdictions are involved.

VII.4.2. **Harmonisation efforts :**

International initiatives continue to seek greater harmonisation of patent systems, notably through treaties such as the PCT and the TRIPS Agreement, as well as ongoing discussions on a unitary European patent that would offer uniform protection in all EU Member States.

CHAPTER 5.
MANAGEMENT OF PATENT LICENCES AND AGREEMENTS

I. Patent valuation tools :

The value of patents depends not only on the technical robustness of the invention, but also on the effectiveness with which it is communicated to investors and strategic partners. Valuation tools play a very important role in attracting interest, persuading stakeholders and facilitating financial or strategic commitments. The main valuation tools include teasers, pitches, investor presentations and other advanced methods. Each of these tools has a specific function within the valuation process. Table 5.1 provides a structured overview of various internationally recognised patent valuation tools, highlighting their objectives, structures, essential content, and suitable presentation techniques. One of the main advantages of these tools is their ability to quickly capture the interest of stakeholders. Tools such as the teaser are designed to attract attention from the outset. By briefly summarising the strengths of the invention and the opportunities it offers, the teaser arouses curiosity, encouraging people to find out more about the invention. This is crucial in an environment where time and attention are limited resources.These tools are also powerful in convincing potential investors and partners. The pitch and presentations to investors demonstrate the commercial potential and competitive advantages of the invention. By incorporating detailed market analyses, solid financial projections and a clear demonstration of technical viability, these tools build confidence and encourage financial or strategic commitment.Beyond persuasion, valuation tools also facilitate negotiations. The data and projections presented provide a solid basis for discussing the terms of agreements such as licences or partnerships. This ability to structure discussions helps to achieve balanced agreements, maximising the value of patents while minimising the risks.What's more, these tools speed up the decision-making process for investors by providing all the necessary information in a clear and organised way. This can be decisive in obtaining financing or strategic commitments within tight deadlines. They also play a significant role in communicating and promoting the invention, whether at conferences, events or on digital platforms. Their ability to be adapted to different contexts and customised to suit the audience makes them versatile and effective tools. Patent valuation tools are essential for structuring the offer, capturing interest, convincing stakeholders and facilitating negotiations, while effectively promoting the invention. Their strategic use maximises the commercial value of patents and ensures long-term market success.

Table 5.1. Some patent valuation tools.

Valuation tool	Objective	Structure	Key content	Presentation techniques
Teaser	Capturing interest quickly and generate discussion.	Concise summary and attractive visuals.	Summary of the invention Market opportunities Value proposition Call to action	Attractive visual design Clarity and simplicity Use of non technique.
Pitch	Quickly convince the audience in less than 10 minutes.	Introduction Problem/Soluti on Market Business model Conclusion	Problem and solution Target market and opportunities Competitive advantages Call to action	Engaging narrative Use of striking visuals Dynamism and enthusiasm.
Presentations to Investors	Provide full details to commit the investment.	Introduction Market analysis Marketing strategy Projections financial Roadmap	Detailed market analysis Financial projections Marketing strategy Call for investment	Demonstratio n live technology Precise answers to questions Contingency plan.
Executive Summary	Offer a more detailed overview more detailed than the teaser, but less long than a full presentation.	Overall summary Value proposition Strategy Call to action	Overview of the invention Value proposition Marketing strategy Call to action	Clear presentation Use of graphics to illustrate the key points.
Business plan	Describe in detail the business strategy, financial projections and planning operational.	Introduction Market analysis Marketing strategy Financial projections Operational plan	Comprehensiv e market analysis Business model Detailed operational plan Financial projections	Use of tables and graphics Detailed, organised presentation.
Product demo	Illustrate the invention and its potential in concrete terms.	Live demonstration or video Presentation of the product and its functions	Presentation of the prototype or final product Key features Competitive advantages	Interactive demonstratio nClear explanation of how it works Answers to questions techniques.

II. Types of licences and intellectual property agreements :

II.1. Definition:

In a licence agreement, the holder of an intellectual property right in a technology grants a third party, under specific conditions, the right to exploit that technology in a given territory and for a specified period. Unlike a sale or assignment, the licensor retains the intellectual property of the technology and grants only a limited right of exploitation. This enables the licensor to grant other licences to different licensees for other geographical markets or business sectors, thereby increasing its sources of revenue. In addition, the conclusion of a licensing agreement often creates a lasting relationship between licensor and licensee, aimed at the efficient and profitable exploitation of the technology. If the collaboration is successful, both parties may see their revenues increase progressively as a result of sales of products using the technology. The legal and practical implications of licence agreements differ considerably from those of sales or assignments, and the commercial objectives pursued require particular attention.

II.2. Different types of licence agreements :

II.2.1. Full or partial licences :

• **Full licence:** The licensee is authorised to carry out all the types of operation stipulated (manufacture, sale and/or use) for all possible applications (medical, agricultural, maritime, etc.).

• **Partial licence:** The licensee is only authorised to exploit the technology for certain types of use or for certain applications. For example, a partial licence for a paint manufacturing process patent could be granted solely for the automotive sheet metal industry, excluding its use in the construction sector. In this case, the parties must define precisely the scope of the applications and the authorised methods of exploitation.

II.2.2. Exclusive or non-exclusive licences :

• **Exclusive licence:** The patent holder undertakes not to grant any other licence for the same technology, application or territory, on pain of incurring contractual liability.

• **Non-exclusive licence:** The patent holder reserves the right to grant other licences for the same technology, application or territory.

II.3. Differences between a licence contract and an assignment contract/

A licence agreement and an assignment agreement are two types of contractual agreement relating to intellectual property, such as patents, trademarks or copyright. However, the rights they grant differ depending on their subject matter.

- **Licence agreement:** The owner of the intellectual property grants a licensee the right to use its property under certain conditions and limitations. The licensee may use the intellectual property for a specific period and in a given territory, but does not become the owner of the intellectual property itself.
- **Assignment contract:** The owner of the intellectual property transfers its rights permanently to another person or company. The transferee acquires full ownership of the intellectual rights and may use, sell or license these rights to others without the consent of the transferor.

III. Negotiating and managing licences :

III.1. Definition:

Negotiating a technology licensing contract is the art of reaching an agreement in which the licensor grants and the licensee acquires the right to exploit a technology on agreed terms. The aim is to create a satisfactory and beneficial relationship for both parties. The success of the negotiation depends on mutual recognition of each party's contributions and an understanding of their respective needs and expectations.

III.2. Phases of the negotiation process :

- **Preparation:** Preparation is crucial to the success of the negotiation. At this stage, the parties have decided that licensing is the best way to achieve their commercial objectives, and that the other party is the most appropriate partner. Good preparation involves defining expectations, appointing a lead negotiator, and drafting a document listing the key commercial issues.
- **Discussion:** During this phase, the licensor highlights the advantages of its technology, while the licensee evaluates the information provided under the confidentiality agreement. Discussions remain general, focusing on the potential benefits of the technology and its relevance to the licensee's objectives.
- **Proposals and haggling:** The parties discuss key commercial terms and aspects of their

potential relationship. This phase often involves give and take, with the aim of reaching a mutually satisfactory agreement.

III.3. Guiding principles for negotiation :

- Seek a mutually beneficial outcome.
- Fix the extremes of its position.
- Aiming high while remaining credible.
- Make concessions on variables that are inexpensive but valuable to the other party.

III.4. Licence agreement and main clauses :

The essential clauses of the licence agreement include legal aspects, the purpose of the agreement, territory and exclusivity, ownership and exploitation of improvements, confidentiality and non-competition, royalties and lump sums, and the duration of the agreement. A well-drafted contract ensures a lasting partnership and minimises the risks for both parties.

IV. Intellectual property agreement :

IV.1. Definition:

An intellectual property agreement is a formal contract that sets out the terms and obligations of the parties regarding the use, commercialisation and protection of intellectual property rights. Such agreements may include licensing agreements, assignments, non-disclosure agreements, joint research and development agreements, and other forms of collaboration that involve the sharing or exploitation of intellectual property.

IV.2. Types of agreement :

There are several types of intellectual property agreement, depending on the commercial objective and the needs of the parties:

- **Licence Agreement:** Agreement whereby one party (the licensor) grants another (the licensee) a right to use specific intellectual property under certain conditions.
- **Assignment Agreement:** Permanent transfer of intellectual property rights from one party to another.

• **Non-Disclosure Agreement (NDA) :** Agreement designed to protect confidential information exchanged between the parties.

• **Joint Research and Development Agreement:** Collaboration between parties to develop new technologies, with sharing of the intellectual property rights resulting from the research.

IV.3. Importance of intellectual property agreements :

These agreements are crucial for protecting innovations, clarifying the rights and obligations of the parties, and ensuring the secure and advantageous commercial exploitation of technologies. They establish the legal bases needed to avoid disputes and ensure that each party benefits fairly from the agreement.

IV.4. Key clauses in agreements

Typical clauses in intellectual property agreements include :

• **Definition of rights:** Specification of the rights granted or transferred.

• **Duration and renewal :** Period during which the agreement is in force and conditions of renewal.

• **Confidentiality:** Protection of sensitive information exchanged.

• **Royalties and payments :** Financial terms relating to the use of intellectual property.

• **Dispute resolution:** Procedures to be followed in the event of disagreement or breach of the agreement.

CHAPTER 6

VALUING PATENTS: SOCIETAL, POLITICAL AND ECONOMIC ISSUES

I. The societal challenges of patent exploitation :

The valuation of patents is much more than a simple economic process aimed at maximising the profits of inventors or companies. It is part of a broader framework in which social, ethical and even cultural implications play a central role. Patents, as instruments of intellectual property protection, give holders a temporary monopoly on the exploitation of their inventions. This monopoly, while essential for encouraging innovation, raises crucial questions about its impact on society as a whole.One of the most significant societal issues in the use of patents is access to technology and innovation. Patents can create barriers to access to new technologies, particularly in critical areas such as health, the environment and agriculture. In the pharmaceutical sector, for example, patents on essential medicines can limit people's access to necessary treatments, especially in developing countries. This situation raises major ethical dilemmas: how to reconcile the need to reward innovators while ensuring that technological advances benefit the greatest number of people? The balance between the right of inventors to be remunerated for their discoveries and the right of people to access these innovations is a delicate one, and requires careful consideration. In addition, the exploitation of patents can exacerbate economic and social inequalities. Developed countries, which have greater financial, technological and legal resources, are often better equipped to take advantage of patent systems. This enables them to strengthen their technological lead, while developing countries, which do not have the same resources, can find themselves marginalised. This dynamic reinforces global economic disparities, creating a technological gap between rich and poor nations. Gaps in access to patents can also limit the local development of technologies adapted to the specific needs of developing countries, reinforcing their dependence on foreign technologies.The use of patents also raises questions of social justice. In many cases, patents are filed by large companies which, although they invest in research and development, often rely on traditional knowledge or on discoveries resulting from public research. Aboriginal communities, for example, possess ancestral knowledge that can be exploited for commercial purposes. When this knowledge is patented without prior consent or without the communities concerned benefiting, this can lead to forms of exploitation and spoliation. Patent protection must therefore be designed to respect the rights of local communities and ensure a fair distribution of benefits. The use of patents also poses challenges in terms of sustainability and environmental responsibility. Patented technological innovations can have considerable environmental impacts. In agriculture, for example, patents on genetically modified seeds (GMOs) have given rise to debates about their long-term effects on biodiversity and food security. Similarly, in the industrial sector, certain patented technologies can contribute to pollution or the depletion of resources. natural resources. It is therefore crucial that patent valuation incorporates sustainability considerations, by encouraging innovations that not only offer economic benefits, but are also respectful of the environment and natural resources. The role of governments and international organisations is essential in tackling these societal challenges. Public policy must seek to balance the interests of inventors with those of society as a whole. This can be achieved through compulsory licensing mechanisms in cases where patents are deemed essential to the public interest, as in the case of life-saving medicines. In addition, incentives can be put in place to encourage patents in the areas of sustainable innovation and green technology. Governments can also support technology transfer initiatives to developing countries, in order to reduce global inequalities in access to innovation.International organisations, such as the World Intellectual Property Organisation (WIPO), play a key role in defining the rules governing patents worldwide. They can promote fairer patent practices and encourage collaboration between countries to tackle global challenges such as climate change or pandemics. The international legal framework must evolve to take account of the new challenges posed by the use of patents in an increasingly interconnected world. It is also crucial to promote an inclusive innovation culture. This means that patenting processes must be transparent and accessible to all innovation stakeholders, including small businesses, independent researchers and local communities. Education and awareness of the patent system are essential to enable more people to protect and exploit their innovations. In addition, public-private partnerships can play a key role in developing innovations that meet social and environmental needs.Finally, it is important to emphasise that the exploitation of patents should be seen not only as a profit-making tool, but also as a lever for social progress. Patented innovations have the potential to transform lives, solve global problems and contribute to sustainable development. However, for these innovations to reach their full potential, they must be integrated into an ethical and societal framework that values not only the inventors, but society as a whole.

II. The political stakes of patent valuation :

II.1. Government strategies for patent exploitation :

Morocco has put in place a set of bold policies aimed at promoting the use of patents and stimulating entrepreneurship. These measures focus on supporting research and development, promoting technology transfer, financing patents, and training and raising awareness among stakeholders. The policies The public sector plays an essential role in facilitating the transformation of innovations into marketable products, while strengthening the competitiveness of Moroccan companies on the global market.

II.2. Support for research, development and technology transfer:

Technological progress and the sharing of knowledge are fundamental drivers of innovation. Morocco actively supports these dynamics by funding research and development projects, while facilitating technology transfer between academic institutions and businesses. Programmes such as the ESRI 2030 PACT bear witness to this ambition, by strengthening research capacities and ensuring that scientific results are exploited through the filing and exploitation of patents.

II.3. Financing and development of patents :

To encourage the valorisation of patents, Morocco offers various financing mechanisms through institutions such as ANPI, the Hassan II Fund, and the Centre for Research and Technological Innovation (CRIT). These initiatives, reinforced by OMPIC's IP Marketplace, enable inventors and companies to access financial resources and maximise the management of their intellectual assets. The increase in funding programmes acts as a driver for the commercialisation of patented innovations.

II.4. Intellectual property training and awareness :

Training and awareness-raising are crucial to optimising the use of patents. Morocco is implementing educational programmes to better inform innovators about the patent filing process and the importance of protecting intellectual property. These initiatives aim to strengthen understanding of intellectual property rights, stimulate innovation and promote a culture of respect for patents. Thanks to increased awareness, economic players are better prepared to protect their inventions and fully exploit their innovations on the market.

III. The economics of patents :

Patents play a central role in the global economy as instruments for protecting innovation and intellectual property. They enable inventors and companies to protect their inventions against unauthorised use, while offering them the opportunity to derive economic benefit from their creative work. The economic implications of patents are wide-ranging and touch on several aspects, from encouraging innovation to improving the competitiveness of businesses, not to mention their role in the economic growth of nations. In this context, it is essential to understand how patents influence the economy and the various challenges and opportunities they present.

III.1. Patents as an incentive to innovation :

One of the main economic challenges of patents is their role in encouraging innovation. By granting a temporary monopoly on the exploitation of an invention, patents offer inventors the opportunity to recoup their research and development (R&D) costs and make a profit. Without the protection offered by patents, companies and inventors could be discouraged from investing in innovation, for fear that their ideas would be copied and exploited by competitors without compensation. Not only does this mechanism encourage innovation, it also helps to maintain a constantly evolving economic dynamic, encouraging the development of new technologies and solutions.

Patents, as economic incentives, are particularly important in R&D-intensive sectors such as pharmaceuticals, biotechnology and electronics. In these industries, the costs of developing new products can be extremely high, and patent protection is often essential to justify the investment. For example, the development of a new drug can cost billions of dollars, and without a patent, the pharmaceutical company may not be able to recoup this investment.

III.2. Business Competitiveness and Competitive Advantage :

Patents are also strategic tools for strengthening a company's competitiveness. Having a well-managed patent portfolio can give a company a significant competitive advantage by enabling it to monopolise the exploitation of certain technologies. This can not only limit direct competition, but also pave the way for additional income through licensing or the sale of patents. Companies often use their patents as leverage in commercial negotiations, whether to secure partnerships, access new markets, or protect their dominant position in an existing market. A classic example is technology companies that use their patents to form strategic alliances or cross-licensing agreements with other companies. This enables them to pool risks and maximise the benefits of exploiting new technologies. Patents can also be used to deter competitors from entering the market. The threat of legal action for patent infringement can be enough to discourage a competitor from launching a similar product, thereby protecting the market share of the company that holds the patent. This barrier to entry is a crucial aspect of commercial strategies in competitive industries.

III.3. Contribution to Economic Growth :

Patents make a direct contribution to a country's economic growth by stimulating innovation and encouraging investment in R&D. Countries with efficient and well-regulated patent systems find it easier to attract foreign investment, as companies are assured that their innovations will be protected. This leads to a where innovation fuels economic growth, which in turn stimulates further innovation.In developing economies, patent protection can play a key role in facilitating technology transfer. Multinational companies are more likely to share their patented technologies with local companies if they are confident that their intellectual property rights will be respected. This transfer of technology can accelerate industrial and economic development, while increasing the competitiveness of local companies in the global marketplace. However, the contribution of patents to economic growth is not without challenges. Economies need to strike a balance between protecting inventors' rights and spreading innovation. Excessive patent protection can hamper competition and slow down technological progress, while insufficient protection can discourage innovation.

IV. Opportunities and challenges of patent exploitation:

IV.1. Opportunities linked to the availability of patents :

The example of Toyota's hydrogen fuel cell patents is a perfect illustration of the nuances and strategic implications behind making patents publicly available. Toyota, having invested massively in the research and development of this technology, has surprised many observers by making its patents accessible. This apparently counter-intuitive move raises questions about the company's motives. Why would a car manufacturer, after allocating colossal resources to an innovation, choose to share its discoveries instead of exploiting them exclusively? Directly exploiting these patents in the field of electric cars seems like a risky option, especially when you consider the words of Tesla CEO Elon Musk, who has described the use of hydrogen as an energy source for vehicles as "the worst solution". Musk points out that the production of hydrogen is energetically costly and that the storage of this gas, due to its low density, also poses serious problems. So while the option of exploiting these patents for electric vehicles is tempting, it doesn't seem to be the most sensible route.However, another approach is emerging: Open Innovation. This concept, which is becoming increasingly popular, proposes using the innovation developed by an entity in contexts or markets other than those initially envisaged. Rather than focusing the use of hydrogen fuel cell technology solely on vehicles, this knowledge could be reused in other sectors or for other industrial applications. In this way, Open Innovation encourages cross-sector collaboration, where ideas and technologies cross traditional boundaries to create new opportunities.

IV.2. **OMPIC and the development of patents in Morocco :**

As part of its drive to promote innovation in Morocco, the Office Marocain de la Propriété Industrielle et Commerciale (OMPIC) has launched a significant initiative: the "IPmarketplace" platform. Launched on National Industry Day, this digital platform aims to promote patents nationwide and establish mutually beneficial partnerships between industrial companies and other economic development players.

"IPmarketplace stands out for its ambition to make industrial property a lever for a more productive, innovative and inclusive Moroccan economy. By offering access to a set of 50 innovative projects based on patents that are free to exploit in Morocco, the platform paves the way for potential investments of 900 million dirhams and the creation of 2,500 jobs. This initiative is a concrete example of how the pooling of intellectual resources can stimulate innovation and economic development.

The success of this initiative lies in OMPIC's ability to align its objectives with market needs. By offering companies the opportunity to exploit existing patents, the platform reduces barriers to entry for new companies and encourages the rapid commercialisation of innovations. This illustrates a pragmatic approach to patent exploitation, where the focus is not only on protecting intellectual property, but also on exploiting it economically.

IV.3. **Challenges and issues in patent valuation:**

While patents are undeniably a driver of innovation, they can also, paradoxically, act as a brake on that same innovation. By granting a temporary monopoly on a technology, patents can limit the use of that technology and, in some cases, lead to higher prices than in a competitive market situation. This paradox is particularly evident in fields where innovation is cumulative, i.e. where each new invention is based on previous discoveries. In such cases, a patent on a basic innovation can delay or complicate the development of new technologies, as subsequent innovators must either wait for the patent to expire or obtain a licence, which is often costly. One historical example is the Wright brothers, whose patent on powered flight slowed the development of aeronautics in the United States. Their competitors, such as Glen Curtiss, were unable to exploit their own innovations immediately because of the Wrights' refusal to license their patent. This situation shows how a poorly balanced patent system can inhibit technological progress.Patents are therefore a double-edged sword: on the one hand, they protect and encourage innovation, but on the other, they can hinder the dissemination and rapid evolution of technologies. The challenge is to strike a balance between these two aspects in order to maximise social well-being.

IV3.1. **New patent challenges :**

Technological developments are creating new challenges for the patent system. The dematerialisation of products and the growing complexity of innovations raise new questions. For example, while abstract ideas and natural phenomena are generally not patentable, advances in biotechnology and computer science have pushed back the limits of what can be protected by patent. There are also international differences, as illustrated by the protection of genetically modified organisms, which has been patentable in the United States since 1980, but was only introduced in France in 2004. Modern products, particularly in electronics, are often based on hundreds or even thousands of patents, creating a complex tangle known as a patent thicket. "patent thicket". For an innovator, navigating this landscape can be difficult, as they have to obtain licences for each component, which can make innovation unprofitable. Large companies circumvent this problem by building up vast patent portfolios and entering into cross-licensing agreements.Another challenge is the emergence of "patent trolls" or NPEs (Non-Practising Entities). These entities do not produce anything themselves, but acquire patents for the sole purpose of collecting royalties or initiating litigation. Their activity can be extremely profitable, but it stifles innovation by diverting resources into costly legal battles.

IV.3.2 Need to develop the system:

Faced with these challenges, many are calling for a reform of the patent system. An efficient system should limit the number of patents granted to innovations that have high innovation costs, low imitation costs and inelastic demand. However, with the growing complexity of products and the increasing number of patent applications, patent offices are under increasing pressure. Congestion in the examination process can lead to patents being granted for innovations that do not meet the criteria of novelty and inventiveness.

The evolution of the patent system must take account of these changing realities if it is to continue to encourage innovation while avoiding unnecessary barriers to technological diffusion. A well thought-out reform could balance inventors' need for protection with the need to foster a competitive and innovative environment.

Printed by Books on Demand GmbH, Norderstedt / Germany